W9-BIU-885

E
Murphy, Stuart J., 1942–
Just enough carrots
                    DOLE

MAY   1998

Oak Park Public Library
Oak Park, Illinois

DEMCO

MathStart™

COMPARING AMOUNTS

Just Enough CARROTS

SUPER STAR MARKET

BY **STUART J. MURPHY**

ILLUSTRATED BY

FRANK REMKIEWICZ

HarperCollins*Publishers*

LEVEL 1

To Amy Edgar—for whom just enough will never be good enough
—S.J.M.

To Kendra—who knows how to count carrots
—F.R.

The illustrations in this book were done in watercolor and Prismacolors on Bristol paper.

HarperCollins®, 🎬®, and MathStart™ are trademarks of HarperCollins Publishers Inc.
For more information about the MathStart Series, please write to HarperCollins
Children's Books, 10 East 53rd Street, New York, NY 10022, or visit our web site at
http://www.harperchildrens.com.

Bugs incorporated in the MathStart series design were painted by Jon Buller.

Just Enough Carrots   Text copyright © 1997 by Stuart J. Murphy   Illustrations copyright
© 1997 by Frank Remkiewicz   Printed in the U.S.A. All rights reserved.

Library of Congress Cataloging-in-Publication Data
Murphy, Stuart J., date
  Just enough carrots / by Stuart J. Murphy ; illustrated by Frank
Remkiewicz.
    p.   cm. — (MathStart)
  "Level 1, comparing amounts."
  Summary: While a bunny and his mother shop in a grocery
store for lunch guests, the reader may count and compare the amounts
of carrots, peanuts, and worms in the grocery carts of other shoppers.
  ISBN 0-06-026778-X. — ISBN 0-06-026779-8 (lib. bdg.)
  ISBN 0-06-446711-2 (pbk.)
  1. Inequalities (Mathematics)—Juvenile literature.   2. Comparison
(Philosophy)—Juvenile literature.   [1. Counting.]   I. Remkiewicz,
Frank , ill.   II. Title.   III. Series.
QA295.M88   1997   96-19495
513.2—dc20   CIP
                                                AC

Typography by Alicia Mikles
1 2 3 4 5 6 7 8 9 10
❖
First Edition

Just Enough CARROTS

SUPER STAR MARKET

OAK PARK PUBLIC LIBRARY
DOLE BRANCH

4

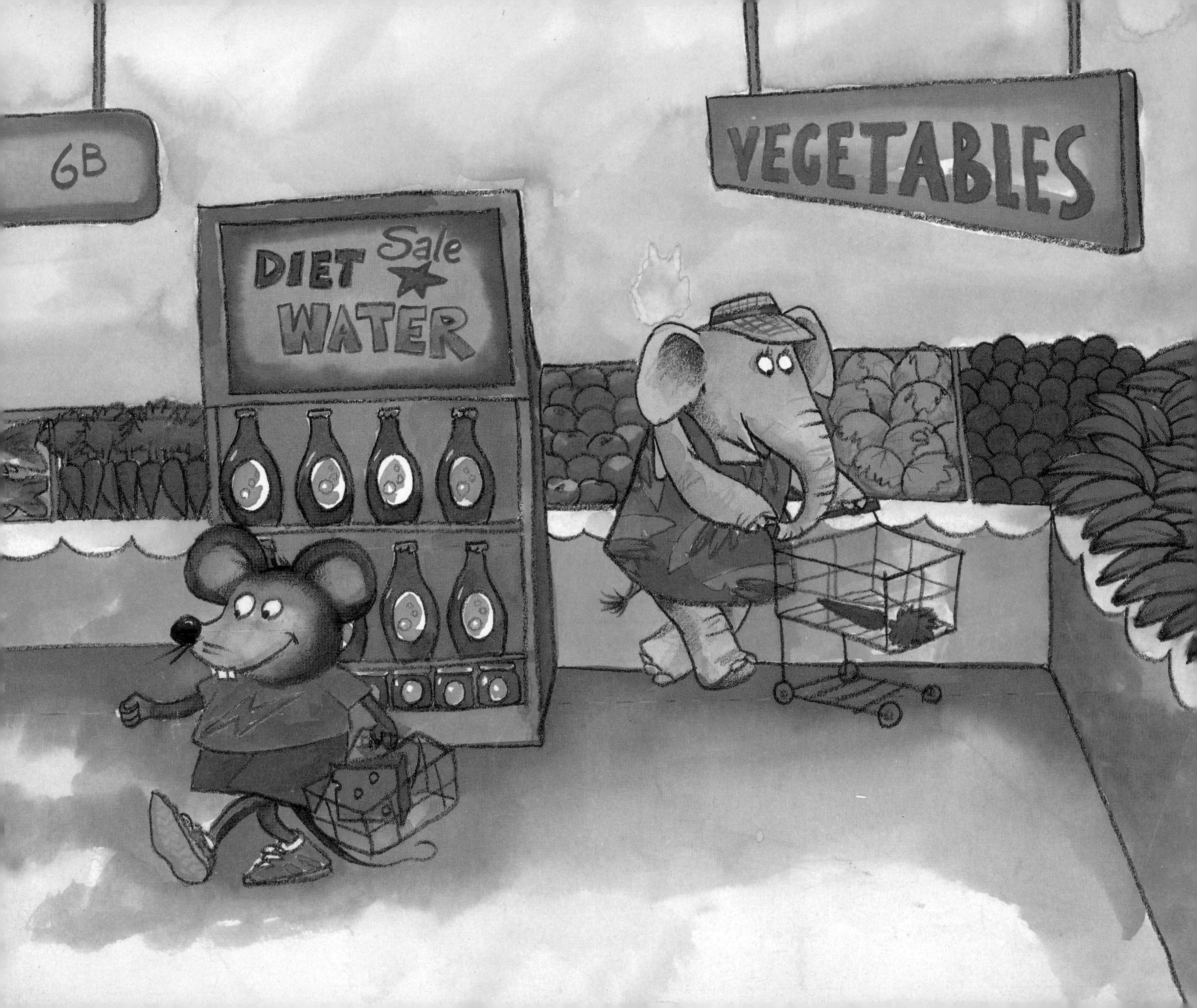
6B
VEGETABLES
DIET Sale WATER

I want some more carrots.
I really like carrots.

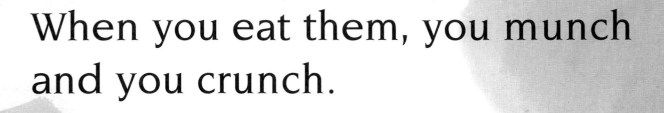

When you eat them, you munch and you crunch.

I know there are others
who have many more carrots.

8

But we have just one little bunch.

Yes, Horse has more carrots,

but Bird has the same amount,

and Elephant has even fewer.

| our carrots | more | same | fewer |

We have too many peanuts.
I don't like to eat peanuts.

First you chomp, then you chew
and you chew.

I'm sure there are others who have fewer peanuts.

14

Why can't we buy just a few?

15

Yes, Squirrel has fewer peanuts,

but Bird has the same amount,

and Elephant has even more.

| our peanuts | fewer | same | more |

Please don't buy any worms.
I really hate worms.

They squish and they squirm
and they crawl.

I'll bet there are others
who have fewer worms.

Let's not buy any at all.

Yes, Elephant has fewer worms,

but Frog has the same amount,

and Bird has even more.

our
worms

fewer

same

more

You don't have to eat peanuts.
You don't have to eat worms.

And we have just enough carrots
to munch.

I bought all of these peanuts

and all of these worms

because Elephant and Bird

are coming for lunch.

Chomp, chomp, and chew.
Squish, squirm, and crawl.

Munch, munch, and crunch—
yummy carrots for lunch!

31

FOR ADULTS AND KIDS

I f you would like to have more fun with the math concepts presented in *Just Enough Carrots*, here are a few suggestions:

- Read the story together and ask the child to describe what is going on in each picture.

- Ask: "Who has more?" "Who has fewer?" "Who has the same amount?"

- Ask questions throughout the story, such as "Would you eat more carrots than the rabbit would eat?" "Would you eat the same amount of worms as the bird would eat?"

- Encourage the child to tell the story alone using the math vocabulary: "more," "fewer," "same."

- Look at things in the real world, such as objects around the house. In the bathroom, are there more toothbrushes than there are bars of soap? Are there the same number of bath towels as there are washcloths? Are there fewer of some things? Try doing the same thing in the kitchen or the bedroom.

- Gather together some items—toys, plastic spoons, blocks—and ask the child to make piles that are more, fewer, and the same.

- Look around you at the library, in a restaurant, or in a department store. Who has more books than you at the library? Can you find a table with the same number of people as at your table?

Following are some activities that will help you extend the concepts presented in *Just Enough Carrots* into a child's everyday life.

At the Playground: Count the number of boys and girls on the swings or lined up for the slide. Are there more girls than boys? The same? Fewer? Are there more children on the swings than on the slide?

Having a Snack: Invite some friends to the kitchen for a snack. Hand out different amounts of cookies, grapes, or peanuts. Who has more? Fewer? The same? Who wants more?

Comparing Buttons: Compare the number of buttons you and two friends have on your clothing. Decide who has "some" and who has "more," "same," or "fewer."

Sorting Crayons: Sort through a box of crayons by color. Choose one color to be "same" and compare the number of crayons in other colors to your chosen color. For instance, you might compare the number of blue crayons to the number of green or brown crayons.

The following books include some of the same concepts that are presented in *Just Enough Carrots*:

- PLANTING A RAINBOW by Lois Ehlert

- HOW MANY TWOS? by Judy Hindley

- MY KITCHEN by Harlow Rockwell